Y0-BPZ-868

The Heat Files

Discovery Channel School Science Collections

© 2000 by Discovery Communications, Inc. All rights reserved under International and Pan-American Copyright Conventions. No part of this book may be reproduced in any form or by any electronic or mechanical means, including information storage devices or systems, without prior written permission from the publisher.

For information regarding permission, write to Discovery Channel School, 7700 Wisconsin Avenue, Bethesda, MD 20814.
Printed in the USA ISBN: 1-56331-008-0

1 2 3 4 5 6 7 8 9 10 PO 06 05 04 03 02 00

Discovery Communications, Inc., produces high-quality television programming, interactive media, books, films, and consumer products. Discovery Networks, a division of Discovery Communications, Inc., operates and manages Discovery Channel, TLC, Animal Planet, Discovery Health Channel, and Travel Channel.

Writers: Jackie Ball, Stephen Currie, Julie Danneberg, Kathy Feeley, Katie King, Susan Lewis, Monique Peterson, Anna Prokos, Jim Scorzelli.
Editor: Katie King. **Photographs**: Cover & p. 2, firefighter w/ flames, ©PhotoDisc; p. 8, James Joule, Mary Evans/Photo Researchers, Inc.; p. 9, Lord Kelvin, Brown Brothers, Ltd.; p. 10, Galileo's thermoscope, ©David Lees/CORBIS; Galileo, ©ArtToday; p. 13, man in furo bath, ©Bob Krist/CORBIS; p. 14, dormouse, ©Suzanne Danegger/Photo Researchers, Inc.; Australian burrowing frog, Jim Merli/Visuals Unlimited; p. 15, woolly bear caterpillar, ©Charles W. Mann/Photo Researchers, Inc.; emperor penguin, ©Fritz Polking/Visuals Unlimited; Polaris fritillary butterfly, ©Cerone/SuperStock, Inc.; p. 16, Bill Phillips, ©Robert Rathe/NIST; p. 17, NIST laboratory, © National Institute of Standards and Technology; Mag-Lev train, Takeshi Takahara; MRI scan, Simon Fraser/Dept. of Neuroradiology, Newcastle General Hospital/Science Photo Library/Photo Researchers, Inc.; p. 26, fireman with fire in background, © PhotoDisc; Bristol Lake, Mojave Desert, ©Renee Lynn; p. 27, thermograph of house, NASA/Science Source/Photo Researchers, Inc.; hot spring in Reykjavik, Iceland, ©Superstock; p. 30, Ice Hotel, Sweden, Chad Ehlers/Stone; p. 31, hot coals, ©Marco Modic/CORBIS; p. 31 astronaut, ©PhotoDisc; all other photos ©Corel.

The Heat Files

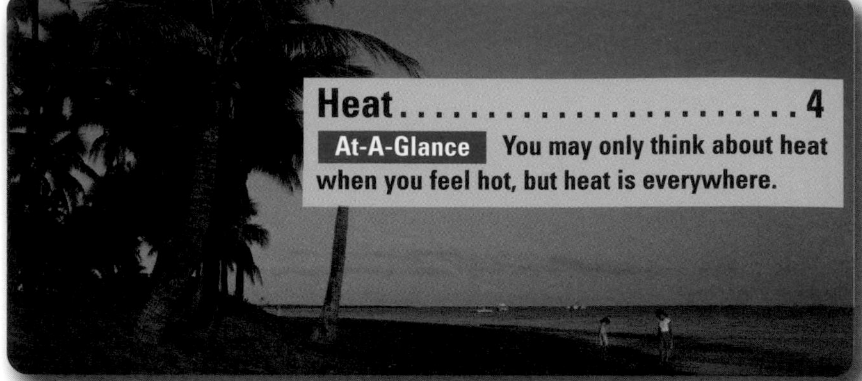

Heat 4
At-A-Glance You may only think about heat when you feel hot, but heat is everywhere.

Hot Stuff. 6
Q & A An infrared ray gives us a glimpse inside the origins of heat.

Heat Can Be Cool: Ask Kelvin and Joule 8
Heroes Learn what James Joule had to do with the joule, and what Lord Kelvin had to do with the kelvin.

Measure for Measure. 10
Almanac How do you measure heat? Very carefully!

Heat Is Happening. 12
Scrapbook Heat makes things happen: Learn how heat is a large part of your everyday life.

Bundle Up! 14
Amazing But True Animals have special ways of regulating their own body heat.

The Quest for Cold! 16
Scientist's Notebook Travel inside a scientific lab to see how physicist Bill Phillips is trying to reach a near-impossible goal: absolute zero.

Turn Up the Heat

Heat. We know it so well we hardly ever think about it. Yet all living things need heat to survive. Heat is everywhere and it affects everything: where we live, work, and build our towns and cities. It affects our weather and sometimes our moods.

We can detect the transfer of heat in our everyday lives: an overheated building, a kettle of water on the stove, or in a natural event like a thunderstorm.

In SIZZLE, Discovery Channel takes you on a tour of heat: where and how it moves, how it affects us, how we use and measure it, and why we simply cannot live without it. But what is heat really? Let SIZZLE tell you everything about heat you've been burning to know!

Warming the Globe 18
Map Find out why some places on Earth are hotter than others—and it's not just about the weather.

Turning up the Heat 20
Timeline Find out what happens to a piece of pizza after it's cooked in a conventional oven and in a microwave oven.

Desert Jaws 22
Eyewitness Account William Lewis Manly and several others stumbled into Death Valley in 1849. This is their story of survival.

Skintight. 24
Virtual Voyage Get into a skier's skin to find out the real story of hot and cold.

The Power of Heat 26
Picture This The effects of heat can be pretty wild and unexpected. See how it works.

Hot Shot 28
Solve-It-Yourself Mystery Summer and Kelvin want to know why Uncle Sonny is being so mysterious about his barn. Infrared film may give them a few answers!

Think Thermal. 30
Fun & Fantastic You can't take the heat anymore?! Heat records, jokes, and more.

Hot pizza delivery at its best: See page 20.

Final Project

Keeping Warm and Keeping Cool . . . 32
Your World, Your Turn Research and design energy-efficient structures to see what keeps them warm and what keeps them cool.

AT-A-GLANCE

Heat

"It's 12:00 and outside it's ninety-two degrees Fahrenheit. Relative humidity is seventy-five percent. Today's forecast calls for a mix of hazy sun and clouds . . ."

Whhhhhew. It's hot! Your radio is blaring while you lounge on the beach, catching some rays from the Sun. When you squint and look down the beach you see waves of heat floating above the hot sand. You can practically feel the earth sizzle. Your skin is absorbing the radiant energy from the Sun, making you hotter. Your skin begins to sweat to help cool you off by radiating some of the heat away from your body. Lying in the shade of your beach umbrella also helps to cool you down. The umbrella absorbs the Sun's radiant energy and conducts heat through the metal rod into the sand, just as a lightning rod deflects a bolt of lightning from a house.

So, is it temperature that makes you so hot? Not exactly. Temperature is only the measure of heat. All objects have heat, some more and some less. The sand absorbs the energy from the Sun and so does the ocean water, but each does so in a different way. The sand absorbs energy more quickly while bodies of water absorb it more slowly. This explains why the sand is extremely hot at noon while the ocean water is relatively cool. Heat is a form of energy that an object has when its atoms and molecules move.

Heat is all around you; it's transferred in three ways.

Radiation

The energy from the Sun reaches you by electromagnetic radiation, especially the infrared variety. The Sun's energy you get at the beach is largely infrared radiation. Your body absorbs the radiation and transforms it to heat energy. Not all the energy is absorbed, though. When you do absorb this heat, the energy of your body increases and you feel hotter.

Conduction

The heat in your beach umbrella rod travels by conduction; the molecules inside the rod increase their energy of motion. As you sit on the beach, the sand absorbs sunlight and is also a pretty good conductor of heat. If one part of the sand is heated by the Sun, a direct source of heat, the neighboring parts will also become heated. All of the sand feels hot.

Convection

Hot air rises, right? Right. The heat of one area of air is physically moved by currents. This is convection. When the air around you is heated, the lighter, warmer air rises, while the colder, denser air sinks. The uneven distribution of heat creates more air currents. Remember: Heat is always present.

Q & A

Hot Stuff

An infrared ray with a lot to say

Q: We're talking with an infrared ray about heat. At least we think we're talking to an infrared ray, although we can't be sure because we can't see it. Are you out there, Ray?

A: Sure. Trust me, I am here, and I can make things HOT—hotter than the rays of the Sun that you *can* see.

Q: Yes, we've heard about how hot you can make things. In fact, some say you're an authority on heat.

A: Well, I hate to boast, but I guess some might call me that. It's all in a day's work—or a year's work—or a few million years' work.

Q: What do you mean by that?

A: I mean, it wasn't easy or quick becoming such an expert. The process started long, long ago and far, far away. Ninety-three million miles away, to be exact. Or 150 million kilometers, if you'd rather.

Q: Ninety-three million miles? But that sounds like the distance between Earth and . . .

A: The Sun. Right you are. The Sun, that big glowing ball of gas, Earth's nearest star, is responsible for nearly all the energy on your planet. You might call it the star of Earth's energy show.

Q: But how can the Sun be responsible for that much of Earth's energy? It's so far away.

A: In space terms, it's not far at all. Besides, even though it's millions of miles away, it's very, very hot. The Sun is really a gigantic furnace, with the elements hydrogen and helium at its core reacting in a way that releases lots of energy.

Q: How hot does it get?

A: Every second the same amount of energy is released as millions of hydrogen bombs going off at once. It's 27,000,000°F (15,000,000°C) in the core.

Q: That's hot stuff, all right . . .

A: It happens when heat is transferred from a material with a high temperature to a material with a lower temperature. The unbelievably hot stuff in the Sun's core has to move out to a cooler part of the Sun.

Q: Which is where?

A: Right outside the core. But that area is so thick it takes millions of years for the heat to pass through it.

Q: Then does the energy come down to Earth?

A: Not yet. It passes through another zone, where it's a "cool" seven million degrees Fahrenheit. Then it leaves the Sun.

Q: Then does heat *finally* get transferred?

A: Yes! And that's when it's transformed into light rays. But the rays have different wavelengths. As an infrared ray, I'm longer than a ray of visible light—too long for human eyes to see. And microwaves and radio waves are radiant energy rays. But that's not the end of the story.

Q: No?

A: Nope. Because the radiant energy is absorbed by things on Earth and changes back to thermal energy. Hold your hand to the Sun; it gets hot, it absorbs radiant energy, and that energy is transformed back to thermal energy. That's the way the Sun heats up things on Earth.

Q: Very interesting. But just one thing: If the Sun keeps leaking light waves as you describe, isn't it in danger of . . . burning out, like a used light bulb?

A: Oh, the Sun will stay hard at work for another five billion years, more or less.

Q: And then what will happen?

A: Scientists say the Sun will start to fuse helium into heavier elements. And that will make the Sun swell up. It may get so big, it will swallow up the earth. After a billion years, the Sun will collapse into a white dwarf star—and it will take another trillion years for this great ball of fire to cool off.

Q: Wow! What will happen to you then, Ray?

A: At that point, I guess it will be time to wave good-bye.

Activity

FEEL THE HEAT **Explore the heat-generating properties of radiant energy. You'll need two glasses, two thermometers, and a magnifying glass.**

1. **Fill both glasses halfway with water.**
2. **Place a thermometer in each glass. Make sure both water samples are the same temperature.**
3. **Place the glasses in sunlight. Record any temperature increases.**
4. **Focus the light from the magnifying glass onto one of the water samples. Note any temperature increases.**

How does the increase in heat in the water relate to the amount of radiant energy from the Sun?

HEAT CAN BE

England, 19th Century

You may not know Gabriel Fahrenheit and Anders Celsius, but you're probably familiar with the temperature scales that bear their names. You've probably heard of the watt, named for the scientist James Watt, and you can likely guess who gave his name to the newton, a measure of force.

Not all measurements are as familiar. Neither are all scientists. Take the joule, named after James Joule, or the Kelvin scale, after Lord Kelvin. These scientists made important contributions to thermodynamics, or the study of changes in heat.

Born in 1818, James Joule made his living brewing beer, but in his spare time he conducted scientific experiments involving heat and electricity. According to one story, he was so intrigued by his studies that he spent part of his honeymoon in Switzerland measuring the temperature of local waterfalls.

Joule knew there was a connection between motion and heat. When motors ran, they produced heat. Joule set out to measure the amount of heat produced by different amounts of motion. He started with a simple relationship: More motion should equal more heat. He reasoned that the water at the bottom of a very high waterfall ought to be warmer than the water at the bottom of a smaller one. When the water had farther to fall, it had more energy, and more energy produced more heat.

As Joule refined his measuring systems, though, he saw that he had made an error. Motion did not just "produce" heat. Instead, the energy of motion and heat were related. When an object moved or an electric motor ran, some motion was converted, or changed, into heat. Gradually, Joule came to an important conclusion. Heat was simply another form of energy. Motion could be changed to heat, and heat could be changed back to motion.

Better yet, Joule realized, no energy disappeared in the process. "Nothing is destroyed, nothing is ever lost," Joule wrote. Energy was conserved, or kept; the total energy in the universe remains the same, regardless of its form. The joule (j for short) is the international unit of energy, named in Joule's honor. A joule is used in all kinds of physics. For example, it takes 4.18 j to raise the temperature of 1 gram of water 1 degree Celsius.

James Joule (1818–89)

Joule measured the temperature of waterfalls to understand heat energy.

COOL

Ask Kelvin and Joule

Lord Kelvin (1824–1907)

LORD KELVIN

William Thomson, who later became known as Lord Kelvin, was the son of a science professor. He entered college at the age of ten. Some stories suggest that he sat in on his father's scientific lectures when he was only eight years old. He used to wave his hand wildly in the air, begging to be called on.

Kelvin is best remembered today for his studies of heat. One of his achievements was establishing a temperature scale based on the idea of absolute zero, the coldest possible temperature. It is the temperature at which all motion ceases, or -273.15°C, (-459.67°F). It is impossible to reach absolute zero. Based on the Kelvin scale, the temperature in your classroom is probably around 294 kelvin (294 K).

Like Joule, Kelvin also made important contributions to thermodynamics. Kelvin reasoned that objects carrying lots of motion or some other kind of energy will leak energy into their surroundings. This can be explained by a bowl of hot soup. The heat will leak from the soup until the soup and the room are the same temperature. The heat will never flow back into the soup on its own because heat flows only in one direction: away from things that are hotter.

Kelvin had agreed with Joule's theory about the conservation of energy, but he took it a step further. Though the amount of energy remained the same, the energy itself had a tendency to spread out. That made it hard to use. We can't convert all the energy from a bouncing ball or an electric motor into usable heat; some heat will always escape before it can be reused.

Joule's and Kelvin's ideas have important implications for the future of the universe. If Kelvin is right, for instance, someday in the far-off future all the usable energy in the universe will have leaked into unusable form. Worried? Don't be. The moment would be billions of years away.

Heat flows only in one direction: away from things that are hotter.

Activity

MELTS IN YOUR CUP **Temperature tells you how hot something is, but not necessarily how long it will stay that way. As a simple example, use two cups, two thermometers, and three ice cubes. Place two ice cubes in one cup and the third cube in the other cup. Place a thermometer in each cup. Ice is water at 32°F (0°C). The ice cubes in the cups should be at the same temperature. Do you think the ice cubes melt at the same rate? Make a prediction, then observe what happens as the cubes melt. Why would one cup of ice melt faster than another? Even though the ice cubes start at the same temperature, what other property needs to be taken into account? Use the results of your observations to write a statement that explains the difference between temperature and heat.**

Almanac

Measure for Measure

Think about this the next time you eat an ice cream cone. The cone feels cold, but it contains heat energy. In fact, everything contains heat energy, including the hottest and coldest objects in the universe. Check out all the ways we measure heat energy.

What's Your Read?

The temperature of an object is a measure of heat energy. We measure it with a thermometer. There are many types, but the mercury thermometer is the most common type.

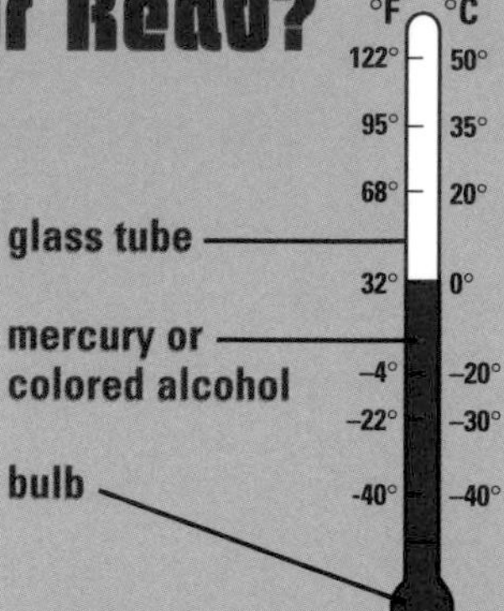

Thermo-WHAT?!

Around 1600 Galileo developed the first primitive thermometer, called a thermoscope (left). He discovered that when materials heat up, they expand, and when they cool down, they contract. Although the thermoscope was not very accurate, it became the basis for all modern thermometers.

To the Point

Some common temperatures to compare

	Fahrenheit	Celsius	Kelvin
Freezing point of water	32°	0°	273 K
Boiling point of water	212°	100°	373 K
Human body	98.6°	37°	310 K
Room temperature	68°	20°	293 K

A Matter of Scale

Thermometers measure temperatures in three scales: Fahrenheit (F), Celsius or Centigrade (C), and Kelvin (K). Unlike the rest of the world, most people in the United States use the Fahrenheit scale to measure body and air temperatures. Although most scientific labs and most school classrooms use the metric system (the Celsius and Kelvin scales), scientists and students occasionally need to use the Fahrenheit scale. Here's how to convert one scale to another.

1. Fahrenheit to Celsius: °C = 5 ÷ 9(°F – 32)
2. Celsius to Fahrenheit: °F = 9 ÷ 5(°C + 32)
3. Celsius to Kelvin: K = °C + 273

Maximum Col

Absolute zero (0 K, -459°F, or -273°C) is as cold as any substance can get. At absolute zero there is no motion. Scientists have not yet achieved this in the lab, though they have created temperatures that are within a fraction of a degree of absolute zero. Compare it with these temperatures:

Liquid hydrogen in the space shuttle's main engine = 20 K (-423°F or -253°C)

Outer space = 3K (-454°F or -270°C)

Blistering Temps

The hottest things are right over your head! Go outside on a cloudless night and look up: The temperatures of stars range from thousands to several billion degrees Celsius. Check out these fiery temperatures.

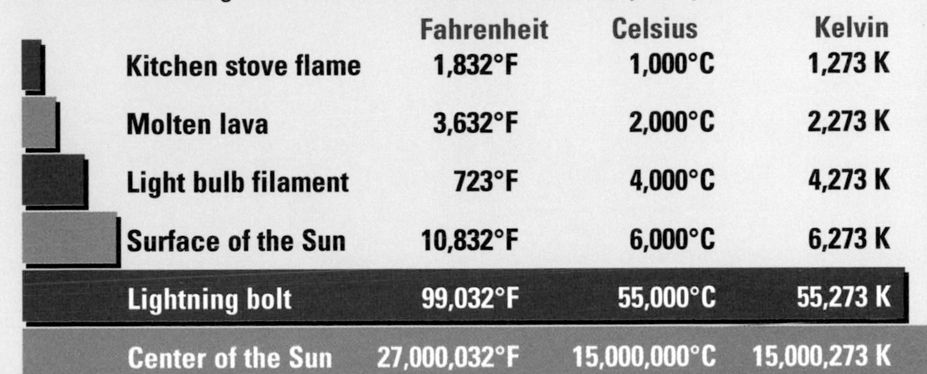

	Fahrenheit	Celsius	Kelvin
Kitchen stove flame	1,832°F	1,000°C	1,273 K
Molten lava	3,632°F	2,000°C	2,273 K
Light bulb filament	723°F	4,000°C	4,273 K
Surface of the Sun	10,832°F	6,000°C	6,273 K
Lightning bolt	99,032°F	55,000°C	55,273 K
Center of the Sun	27,000,032°F	15,000,000°C	15,000,273 K

Cool It

Thermodynamics is the study of the movement of heat.

Thermo = heat

Dynamics = movement

One physical law of heat says that heat can only pass from hotter objects to cooler objects.

Ever notice how much colder you feel when the wind is blowing? Lower temperatures and the wind work together to take away your body heat. Weather reporters call this the windchill factor. If it's 25°F outside, and there's a 30-mph wind, it feels like 10° below zero. The table below shows the air temperature on the top row and the wind speed on the left side. You can find out the windchill factor by plotting where the two scales meet in the center of the table. Bundle up!

AIR TEMPERATURE (°F)

WIND SPEED (MPH)	35	30	25	20	15	10	5	0	-5	-10	-1
10	22	16	10	3	-3	-9	-15	-22	-27	-34	-40
20	12	4	-3	-10	-17	-24	-31	-39	-46	-53	-60
30	6	-2	-10	-18	-25	-33	-41	-49	-56	-64	-71
40	3	-5	-13	-21	-29	-37	-45	-53	-60	-69	-76

Cooks Beware!

Water boils at 212°F, right? Yes, but this changes depending on altitude. At higher altitudes, water boils at lower temperatures. Air pressure is lower at higher altitudes, so recipes must be adjusted. Check out the chart at right.

PLACE	ALTITUDE	BOILING POINT
Mt. Everest, Nepal	29,035 feet	159.8°F
London, England	Sea level	212°F
Miami, Florida	11 feet	210°F
Dead Sea, Israel	-1,296 feet	213.8°F
Denver, Colorado	5,280 feet	203°F
Mt. McKinley, Alaska	20,320 feet	176°F
Death Valley, California	-282 feet	213°F

Note: Temperatures are approximate.

Mount Everest in the Himalayas

Activity

Boil It Down Choose four cities and find data on each city's altitude and the boiling point of water in each city. Create a graph with this data. Then choose four additional cities and try to predict the boiling points in these cities.

HEAT IS HAPPENING

Heat is a kind of energy, and energy makes things happen. Enough heat changes water into steam, batter into cake, and wood into ash. Ever since prehistoric people discovered that fire protected them, kept them warm, and cooked their food, people have been using heat every day of their lives. Our needs for heat aren't much different from those of people living long ago. But technology is more advanced. So instead of fire we use heat-activated motion sensors, hot-water heaters, stoves, and microwave ovens.

Building Smarter

Engineers must build safe roads and buildings. Concrete and metal, two common materials used in buildings and bridges, are good conductors of heat. They expand as heat travels through them. Little gaps are built between sections of highway bridges to leave room for heat expansion. On a hot day, the central span of San Francisco's Golden Gate Bridge (left) is more than 1.5 feet longer than on a cold day.

If the gaps weren't there, the bridge would press in on itself and buckle under the pressure. It's the same thing for railroads: The rails are always positioned with a gap to accommodate the heat from the friction of the rail wheels.

Heaps of Heat

Haystacks, piles of compost, or anything else that decays can suddenly catch on fire on its own. The process of decay causes heat and, if that heat becomes great enough, it can cause such material to burst into flames.

In *The Midwife's Apprentice,* by Karen Cushman, the main character, Beetle, knows a little about heat. A homeless orphan living in medieval times, she is cold, tired, and desperate to find a warm place to sleep.

"When animal droppings and garbage and spoiled straw are piled up in a great heap, the rotting and moiling give forth heat. Usually no one gets close enough to notice because of the stench. But the girl noticed and, on that frosty night, burrowed deep into the warm, rotting muck, heedless of the smell."

Lighten Up

The movement of heat causes some very interesting phenomena. If you've ever experienced a thunderstorm, you know that a lot can happen in a matter of minutes. Heat plays a central role in a thunderstorm. First, a giant discharge of electricity causes lightning. The bolt of lightning heats the air around it to 54,000°F (30,000°C) and forces the air to expand rapidly. This causes a shock wave and then a sound wave—the moment when you hear a clap of thunder.

Enlighten Me

Believe it or not, only 10 percent of a lightbulb's energy is light. Most of the energy a bulb gives off is in the form of heat.

The Laws of Heat

One important physical law of heat is that it always flows away from an area that is hotter—never the other way around. In the play *Arcadia*, by Tom Stoppard, two characters comment on this fact of nature in act 2, scene 7.

Valentine: *Listen—you know your tea's getting cold.*

Hannah: *I like it cold.*

Valentine: *I'm telling you something. Your tea gets cold by itself, it doesn't get hot by itself. Do you think that's odd?*

Hannah: *No.*

Valentine: *Well, it is odd. Heat goes to cold. It's a one-way street. Your tea will end up at room temperature. What's happening to your tea is happening to everything everywhere. The sun and the stars. It'll take a while but we're all going to end up at room temperature.*

You're in Hot Water Now!

In Japan people enjoy the feel-good benefits of a furo, or Japanese-style hot bath. After washing, bathers soak in a big wooden or metal tub, typically using water heated to 110°F (43°C) or hotter. The furo is used in private homes.

A furo is heated by a small electric system or a gas heater. Furos with extra thick wood have greater insulation, so the high temperature can be maintained for a long time. And bathers stay warmer longer!

Activity

Too Hot to Handle

Pour the same amount of hot water into four different-size containers or bowls, each one made of different material, such as Styrofoam, porcelain, metal, and wood. Measure each container's temperature every 5 minutes over a 15-minute period. Which materials are better conductors of heat? Which are better insulators, or retain heat the best?

BUNDLE UP!

You might jokingly call yourself hot-blooded, but it's closer to the truth than you may think!

Humans, as well as birds and mammals, are endotherms, or warm-blooded animals that regulate heat with their own bodies. This helps them keep their body temperatures fairly stable. But reptiles, fish, insects, and amphibians are ectotherms, or cold-blooded animals. Their body temperatures are affected by the temperature of the surrounding air.

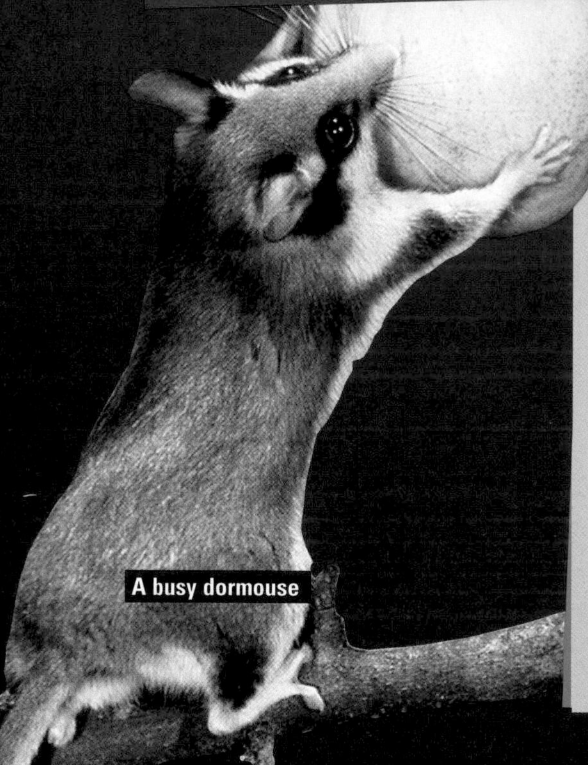

A busy dormouse

Chillin' Out

Some mammals hibernate to survive the winter. Other animals sleep for part of the winter, but that doesn't make them true hibernators. An animal that truly hibernates turns off its body's ability to regulate its temperature, so it drops way below normal. A dormouse (left) is a hibernator. Its body undergoes many complex changes: Respiration slows down to less than one breath a minute, its heart rate grows very weak, and its temperature drops until it is almost as cold as the surrounding air. In this state, the dormouse appears to be dead.

Bears don't hibernate. They rest! Once a bear starts its winter sleep, its temperature drops from 98.6°F (37°C) to about 93°F (34°C), and its heart rate decreases. But bears do not turn off their body's ability to regulate temperature altogether. During this resting period, a bear can be awakened by loud noises. So don't be fooled. Always let sleeping bears lie!

Too Hot to Trot

In some extremely hot places animals—usually cold-blooded—go into estivation, a dormant or sleeplike state. The Australian burrowing frog (right) digs an underground chamber, where it retreats to beat the summer heat. It covers itself with a waterproof skin that looks like a tight plastic bag, to protect against the dry heat outside. As soon as the rains return, the frog removes its skin and emerges from its underground sleep.

Awesome Adaptations

Both cold-blooded and warm-blooded animals have developed different ways of protecting themselves from extremes in temperature.

Arctic woolly bear

1. The arctic woolly bear, a caterpillar, spends most of its life trying not to freeze to death. When fluids freeze they expand, which—along with ice crystals—damages cells and often causes death. The woolly bear produces special chemicals to prevent ice from forming inside its cells.
2. The whale has a layer as thick as 2 feet (.6 m) that protects its insides from icy water.

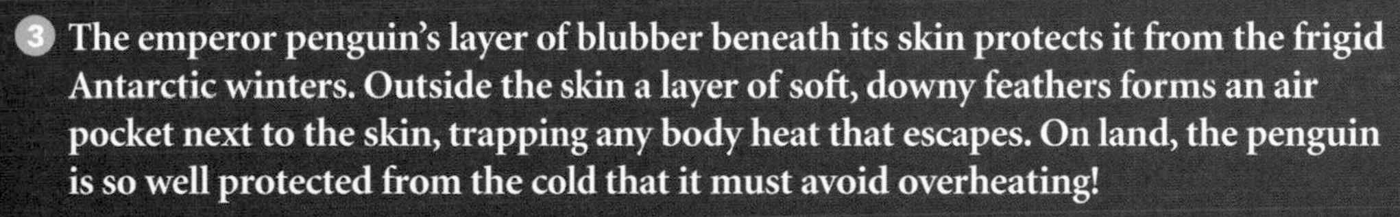

3. The emperor penguin's layer of blubber beneath its skin protects it from the frigid Antarctic winters. Outside the skin a layer of soft, downy feathers forms an air pocket next to the skin, trapping any body heat that escapes. On land, the penguin is so well protected from the cold that it must avoid overheating!
4. A polar bear's fur is really a coat of transparent, hollow hairs that can trap and funnel 95 percent of the Sun's rays. The coat, along with a layer of fat, makes heat loss so minimal that the animal is almost impossible to see with heat-detecting infrared sensors.
5. The Polaris fritillary butterfly turns its wings to the Sun. The wings act like natural solar panels and absorb the Sun's heat.

An emperor penguin with her young

Polaris fritillary butterfly

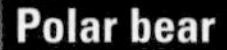
Polar bear

Activity

Hot or Cold? Our skin acts like a crude thermometer because it can sense the difference between hot and cold. But senses can be fooled. See for yourself with this exercise. You'll need: three coffee mugs, water, and a thermometer. Put cold water in one, warm water in another, and room-temperature water in the third. Put two fingers in the warm water and count to 20. Repeat with the room-temperature water. Which feels cooler? Write down your observations. Then record the temperature of each using a thermometer. Record the temperatures after 20 seconds. Now put two fingers from your other hand in the cold water for one minute. Then put those fingers into the room-temperature water. Describe how the water feels. Why do you think this is so?

SCIENTIST'S NOTEBOOK

The Quest for COLD!

Zero degrees Celsius is cold, 0°F is colder, but 0 kelvin—which is equal to -273°C—is the coldest of all. As temperatures decrease, atoms and molecules move more slowly. At 0 K—known as absolute zero—they do not move at all. Zero motion has not yet been achieved in a lab. Physicists try to produce and maintain extremely low temperatures to examine what happens to matter. They're studying superconductivity and superfluidity, conditions at extreme temperatures. Among other developments, this work has led to more accurate and precise timekeeping, faster computers, and safer and more accurate medical technology.

How Cold Is Cold?

For more than 20 years, physicist Bill Phillips has been on a quest to reach absolute zero. At the laboratories of the National Institute of Standards and Technology (NIST) in Gaithersburg, Maryland, he and his colleagues have simulated the coldest temperatures in the universe—temperatures that are only millionths of a degree above absolute zero. The lowest natural temperature—found only in interstellar space—is almost three degrees warmer.

In his laboratory, Phillips uses lasers, or focused beams of light, to slow down gas atoms. Then he "traps" the slowed atoms in electromagnetic fields. A vacuum chamber keeps out other atoms. At room temperature, atoms are difficult to study because they move too quickly. But atoms that have been cooled and trapped are easier to study in detail, even though they remain trapped for only about a minute. Phillips compares the process of slowing, cooling, and trapping atoms to spraying a stream of water at rapidly volleyed tennis balls.

One of the goals of his experiments with super-cold atoms is to improve the accuracy of the official U.S. atomic clock, located in his lab. The atomic clock is the most accurate clock in the world. (Check it out at the Web site www.time.gov.) The uses of this clock are many and varied: maintaining high-speed communications systems, calculating bank transfers, regulating power grids, and synchronizing NASA's interplanetary travel.

Dr. Bill Phillips in his atom-trapping lab

Keepin' His Cool

Minutes, seconds, millionths of degrees . . . Phillips's research requires incredible precision. Times, temperatures, electromagnetic fields, and other conditions must be measured and remeasured with extreme accuracy. Lasers and other equipment must be calibrated, or adjusted, to exact specifications. Because noise can affect the experiments, he must keep the laboratory dark and quiet.

Phillips received the Nobel Prize in physics in 1997 for his work. "Winning the prize has been an enormous honor," says Phillips. "It is thrilling to have your work recognized—and to have made the discoveries in the first place."

What's So Super About Superconductivity?

Research into absolute zero has major implications for technology today. It is already changing our world. Besides the atomic clock, another practical use for absolute zero is superconductivity, the ability of

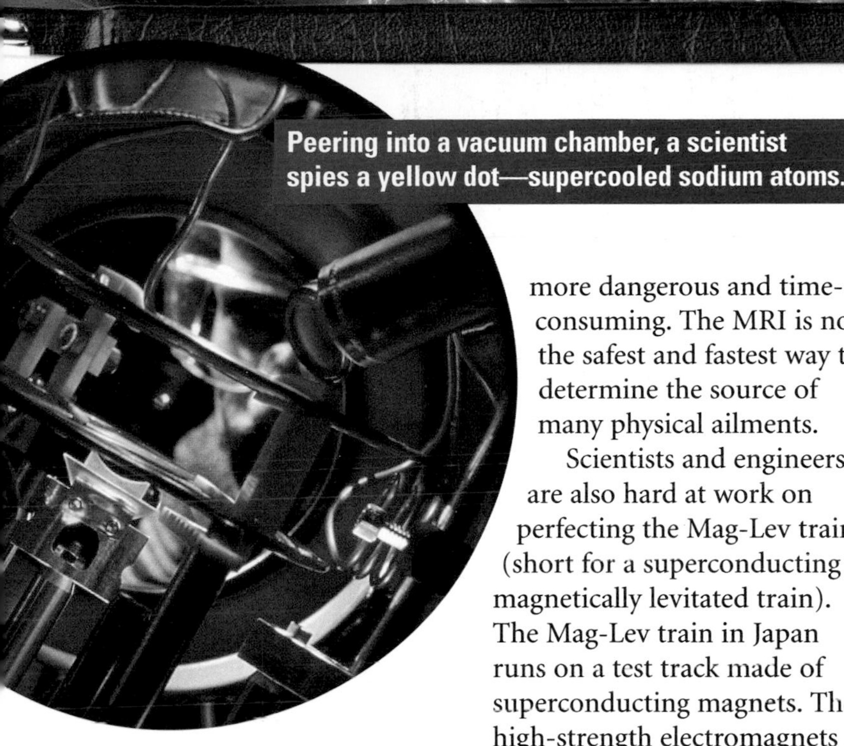
Peering into a vacuum chamber, a scientist spies a yellow dot—supercooled sodium atoms.

A patient undergoes an MRI test.

some materials to lose their electrical resistance at very low temperatures. This means that electrical currents do not lose some of their energy to heat loss. Superconductors are an important part of electromagnets because they can carry the very high electrical currents necessary to produce large magnetic fields without melting. Electromagnets are a key component of Magnetic Resonance Imaging, or MRI. An MRI machine uses radio waves to create a three-dimensional picture of the inside of the human body, a kind of super-detailed X-ray. Doctors use the MRI test to examine soft human cartilage, membranes, and brain tissue. Before the MRI was introduced in the late 1980s, doctors often had to perform invasive exploratory surgery to make a diagnosis, which was far more dangerous and time-consuming. The MRI is now the safest and fastest way to determine the source of many physical ailments.

Scientists and engineers are also hard at work on perfecting the Mag-Lev train (short for a superconducting magnetically levitated train). The Mag-Lev train in Japan runs on a test track made of superconducting magnets. These high-strength electromagnets lift and propel the train along—no more than a few centimeters above a monorail guideway. So the train actually rises, or levitates! One advantage of the Mag-Lev is the lack of wheel-and-rail frictional forces that might someday allow high-speed travel with low environmental impact and minimum maintenance.

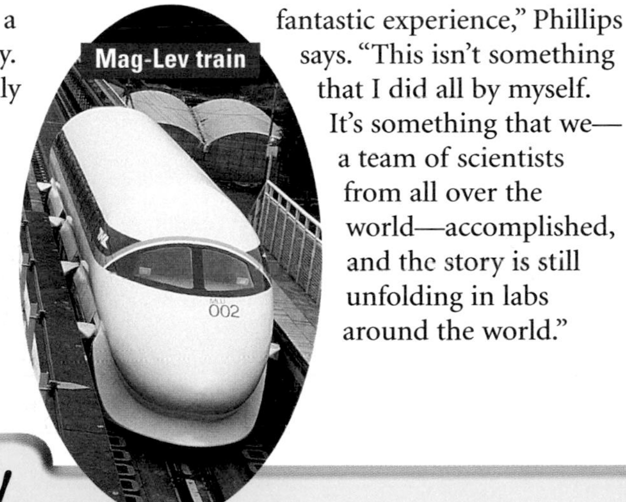

Mag-Lev train

The work that Phillips and others have done in the search for absolute zero (and the creation of a heat-free environment and materials) has had a significant impact on many aspects of daily life and points to new avenues for further research and technological developments. "The quest for absolute zero has been a fantastic experience," Phillips says. "This isn't something that I did all by myself. It's something that we—a team of scientists from all over the world—accomplished, and the story is still unfolding in labs around the world."

Activity

ELECTRIC ENERGY One way to see how heat affects electrical energy is to test an electric circuit at different temperatures. A teacher or parent must help you with this.

Materials:

2 small flashlights, each with new batteries.

1. **Assemble the flashlights and turn them on. Record the time.**
2. **Place one flashlight on a table or desk at room temperature.**
3. **Place the other flashlight in the refrigerator or freezer.**
4. **Record the time it takes for the flashlight at room temperature to dim and eventually lose its power.**

What happens to the flashlight in the refrigerator? What effect does heat have on electric current? What do you think would happen at absolute zero?

Warming the Globe

Earth receives energy from the Sun in the form of radiation. Some of these electromagnetic rays are converted to heat. But only about 43 percent of the solar energy that hits Earth gets to the surface. Another 42 percent is reflected into space, while 15 percent is absorbed by Earth's atmosphere before it can reach the planet's surface.

Heat on Earth's surface and in the air is not distributed equally. Regions receive different amounts of solar radiation, depending on the angle, known as the angle of incidence, at which sunlight hits the globe. The tropics are very hot, in large part because they are near the Equator. Here sunlight strikes the globe from directly overhead and transfers more energy per square meter than if it were to strike at an angle. The less direct the angle of sunlight, the wider the surface area it hits, giving less solar energy.

After the Sun's energy hits the ground, heat moves into the air, largely by convection. It affects the weather and temperatures all around the globe.

Sunlight is not the only source of heat energy. The layers of molten rock and iron beneath Earth's crust are thousands of times hotter per square meter than on the surface. Heat below the crust helps move the planet's tectonic plates and generates geothermal energy.

This map shows how heat makes an impact on the planet.

1 Juan de Fuca Ridge, northeastern Pacific Ocean Heat energy comes up through the earth's surface at hydrothermal vents on the ocean floor. In 1997–98, oceanographers located a group of vents at this point in the earth's crust. Sulfurous gases from the earth's core rise out of the vents, heating ocean water to about 750°F (400°C).

Tropic of Cancer (23.5° N)

Equator

NORTH AMERICA

SOUTH AMERICA

2 Yellowstone National Park, Wyoming This area has more than 300 hundred geysers, hot springs that erupt when water trapped in underground channels becomes superheated by molten rock below the crust. Steam created by the high temperatures forces water out these channels through an opening in the crust. With each eruption, Old Faithful, the most famous of Yellowstone's geysers, releases between 3,700 and 8,400 gallons (14,000 and 32,000 l) of water; it reaches temperatures higher than 204°F (96°C).

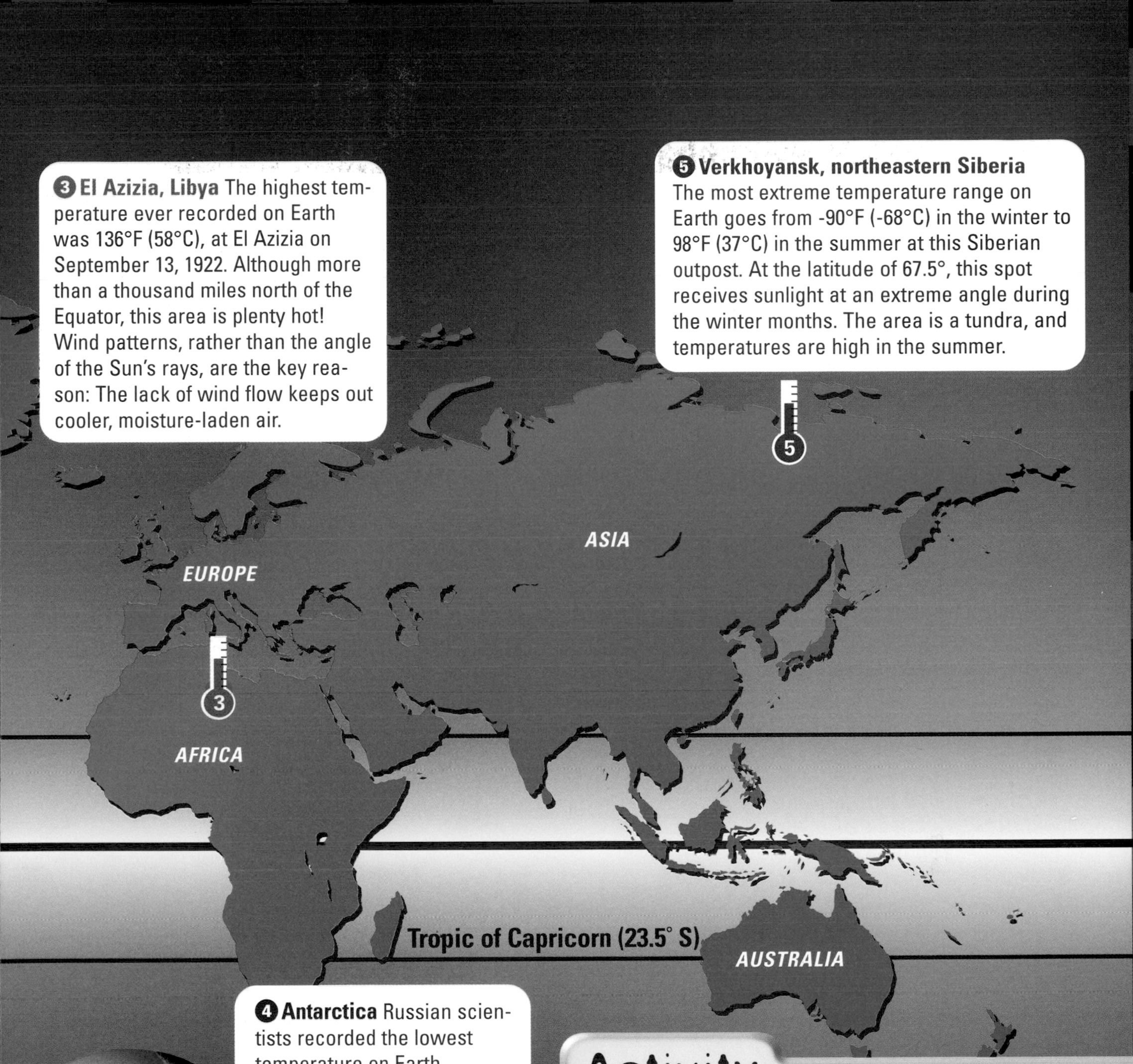

3 El Azizia, Libya The highest temperature ever recorded on Earth was 136°F (58°C), at El Azizia on September 13, 1922. Although more than a thousand miles north of the Equator, this area is plenty hot! Wind patterns, rather than the angle of the Sun's rays, are the key reason: The lack of wind flow keeps out cooler, moisture-laden air.

5 Verkhoyansk, northeastern Siberia The most extreme temperature range on Earth goes from -90°F (-68°C) in the winter to 98°F (37°C) in the summer at this Siberian outpost. At the latitude of 67.5°, this spot receives sunlight at an extreme angle during the winter months. The area is a tundra, and temperatures are high in the summer.

4 Antarctica Russian scientists recorded the lowest temperature on Earth, -128.6°F (-89.2°C), at Vostok II, a Soviet station in Antarctica, on July 21, 1983. As at the North Pole, solar radiation strikes the ground at such an extreme angle that it distributes energy over a large surface area. This means that the solar energy is much less intense per square meter than in areas closer to the Equator.

Activity

A NEW ANGLE Try this model way to see what happens when a beam of light strikes a surface. You'll need a flashlight, two pieces of paper, a pencil, and a ruler. Lay the paper flat and hold the flashlight a few feet up and directly over it. Trace the outline of the beam of light on the paper. Then take the second paper and aim the flashlight at an angle. Trace the outline of the beam of light. Use the ruler to compare the two outlines. Think about what this demonstration shows about sunlight on Earth, and how heat is distributed.

Turning Up the HEAT

TIMELINE

MICROWAVE

Step 1 Start it Up	Step 2 Do the Wave	Step 3 Getting Excited
Place one slice of cold pizza on a plastic or microwave-safe plate and pop it in the microwave oven. Punch in a time and press START.	Microwaves, or waves of electromagnetic energy, are fired into the oven. These waves move across the top of the oven. Radiant energy is either absorbed by the pizza or reflected by the metal walls of the oven until all microwaves are converted to heat.	Different parts of the pizza absorb heat at different rates. Heat begins to spread throughout the pizza.

CONVENTIONAL OVEN

Step 1 Hot Stuff	Step 2 The Heat is On	Step 3 Hot Air Rises
Put a whole pizza on a metal pan or aluminum foil and place it on the oven rack. Unlike the microwave oven, it's not a good idea to put plastic containers in a conventional oven. It gets so hot in there that plastic and other soft materials will quickly melt or burn. Turn the oven up—350°F is about average—and wait.	When you set the oven's temperature, the oven works hard to reach that temperature. Conventional gas and electric ovens work by convection, a process by which heat rises and cold air sinks. The food is heated by exposure to hot air.	Air inside the conventional oven heats up and rises to the top of the oven, just as the microwaves rise to the top of the microwave oven. As the hot air rises, it touches the pizza. Some of that hot air passes onto the pizza to warm it up. The rest of the air gets cooled. That cool air sinks and more heated air moves into the oven to replace it.

There are many ways to heat up leftover pizza. We turned up the heat on a microwave oven and a conventional oven to see how heat moves.

How Hungry Are You?

In a conventional oven, it won't take you any longer to heat up several slices of pizza than it does to heat up one slice. Energy first heats the air inside the oven rather than the pizza. But in a microwave oven, the microwaves heat the pizza, without heating the air first. The microwaves that miss the pizza bounce around until they finally strike it and get absorbed. So it takes longer to microwave one slice of pizza—three to five minutes—than it does to microwave a whole pizza pie.

Step 4
Meltdown

As more microwaves are absorbed by the pizza, more heat is generated. Water begins to evaporate and the cheese begins to melt.

Step 5
Voilà!

More heat is generated and it spreads through the slice in a short time. Some parts of the pizza may heat up faster than other parts. Have you ever noticed that pizza crust gets soggy in the microwave? Here the heat doesn't brown the top of foods to make them crispy. So gulp down that pizza fast—before the cheese cools and becomes too rubbery to chew!

Step 4
You're Getting Warmer

When newly heated air passes over the pizza, only a small amount of heat goes onto the pizza. It takes a long time for enough heated air to circulate through the oven and transfer enough heat to sufficiently warm up the entire pizza.

Step 5
Crusty and Crispy

The heat passes from the top of the pizza through the center and bottom of the pizza to the crust. This process is called conduction. That allows the pizza to become heated and cooked evenly. After about two minutes of 350°F heat, that cold pizza is now hot and bubbly and you've got a crispy crust. Dig in!

Activity

Conduct It Using a microwave oven to heat a slice of pizza relies on radiant energy. Using a conventional oven relies mostly on convection. Here's a way to heat a slice of pizza using conduction.

Place a griddle on the stove top and turn on the burner. Then place the pizza slice on the griddle. How does the heat from the burner warm up the pizza? How does the heat energy get transferred to the pizza slice? How is this an example of conduction?

Desert Jaws

Death Valley, California, 1849

In eastern California on the Nevada border lies a desert known as Death Valley—one of the hottest places on Earth. It was here that William Lewis Manly and his party nearly perished. Manly and the Bennett, Arcane, Earhart, and Wade families stumbled through Death Valley en route to the California goldfields in 1849. Lost, thirsty, hungry, and quickly losing hope, 29-year-old Manly and his friend John Rogers went for help. *Death Valley in '49* is the story of their ordeal as told by Manly.

In Death Valley temperatures are extremely hot and it is hard to find water. It is a dry, desolate place where grass is scarce. Everywhere the Manly party looked, they saw dried lakebeds. A strange land awaited them.

It stood in sharp peaks and was of many colors, some of them so red that the mountain looked red hot, I imagined it to be a true volcanic point, and had never been so near one before, and the most wonderful picture of grand desolation one could ever see.

Their prospects grew worse by the day: There was not enough food or water, even for the children. The oxen were failing. Finally the group sent ahead Manly and John Rogers, "a tough Tennessee man," to find a settlement by foot. The others would wait at a spring.

We, both of us, meditated some over the homes of our fathers, but took new courage in view of the importance of our mission and passed on as fast as we could.

A Little Patch of Ice

Manly and Rogers set out to cross the Panamint Mountains and the Mojave Desert toward the Los Angeles settlement. Early on, thirst set in. They suffered from dehydration.

Our mouths became so dry we had to put a bullet or a small smooth stone in and turn it around with the tongue to induce a flow of saliva.

During the day the sand and rocks got very hot, reflecting sunlight and radiating heat. But once the sun went down, the temperature rapidly dropped. Early one day they found a patch of ice in the sand, and they melted it in their kettle.

I can but think how providential [lucky] it was that we started in the night, for in an hour after the sun had risen that little sheet of ice would have melted and the water sank into the sand.

Death Valley, California, is the lowest surface in the Western Hemisphere. It is 130 miles long and 14 miles across.

After several days of hunger, they scavenged the land for food and shot a black crow. As they approached Wilson's Peak and the San Bernardino Mountains, they knew civilization was in sight. They crossed several miles of snow, and several times they lost the trail. Two weeks later they arrived at a Spanish mission near Los Angeles. Manly managed to purchase three horses and plenty of jerky, cornmeal, and flour by communicating in sign language. Although he was grateful to have met so many friendly people, Manly worried that it might be too late for those he had left behind in Death Valley.

The time for our return was almost up and there was no way of getting back in fifteen days as we had agreed upon, so there was great danger to our people yet.

A Startling Discovery

On their return, Manly's new mule refused to budge.

She must cross this narrow place . . . or be dashed down fifty feet to a certain death. . . . I tell you, friends, it was a trying moment.

Further down the trail, Rogers found a member of their party, Captain Culverwell, who had died on the trail.

He did not look much like a dead man. He lay upon his back with arms extended wide, and his little canteen, lying by his side. . . . How many more bodies should we find? Or should we find the camp deserted?

"The Boys Have Come!"

When they finally arrived at the camp, nearly a month after leaving, the group could hardly believe their eyes. Manly reflected,

Our hearts were first in our mouths, and then the blood all went away and left us almost fainting as we stood and tried to step.

There were eight survivors left of the seventeen or so people who had been left behind to await rescue. Some had gone on; some had died (perhaps of heat stroke). Manly told the group that because the terrain was rough, they would abandon the wagons and walk almost 250 miles. The journey looked grim, yet the survivors felt relieved.

Just as we were ready to leave [camp] . . . and overlooking the scene of so much trial, suffering, and death spoke the thought uppermost saying: "Good bye Death Valley!"

Despite the length and difficulty of their journey, Manly, Rogers, and the rest of the group made it out of the desert alive. Several years later, Manly wrote his account of their trials in Death Valley.

Takin' Some Heat

Your body regulates temperature by either getting rid of or retaining heat. If overheated, the body circulates more blood to the skin to conduct heat out from inside. Sweat glands also help cool the body down. Less blood flows to the skin in the cold, and sweat glands slow down. This allows the skin to conserve heat. Your ideal body temperature is about 98.6°F (37°C).

VIRTUAL VOYAGE

Skintight

When it comes to heat, your body knows how to make it and take it. You manufacture your own heat from the food you eat as well as receiving it from the outside. All the while, your body is constantly regulating its temperature, responding to heat and cold. And the biggest organ in the human body—the skin—plays an enormous part in that regulation. Hold on tight and get under a skier's skin to see for yourself.

6 a.m., somewhere in the American West

The wind rattles the windows of the cabin. Inside, the firewood has burned down and the wood stove is cold. Inside the furnace known as the human body, the fuel has run out, too. For the skier whose body you're tightly wrapped around, last night's dinner was a long time ago. That means little internal fuel to burn, and little resulting heat to conduct outward to you, the skin. Without much heat coming from an external source either, one of the tiny body temperature regulators in the brain slows down the flow of blood out to you. You work with the fat layer under you to hold in whatever heat you can. Your surface becomes lumpy with goosebumps, the body's natural reaction to cold and fear. Goosebumps are caused by tiny muscles around the hair follicles, which contract to lower blood flow and hence conserve heat. Almost your entire surface is covered with hair, although much of it is hardly visible.

7 a.m.

The skier starts the fire in the wood stove. The heat from the hot metal moves through the room. She eats breakfast, refueling her own furnace. You start to feel warmer as circulation of the blood speeds up. The goosebumps are now gone. The skier gets dressed for a day in the wind and the cold. She needs to make sure she's equipped to hold onto all the heat she has. That means dressing in layers. First a thermal underwear top and bottom, next to you. Then a couple of insulating layers—a synthetic blend turtleneck, a fleece vest, and tights. All these layers will trap body heat between them and keep her as warm as possible. Then it's time for ski pants and a parka, and she's off into the bitterly cold day.

11 a.m.

Whoosh! Swish! The skier has been exercising hard all morning, maneuvering her skis over bumpy moguls, slashing through knee-high fields of powder. A second sensor in the brain registers a rise in body temperature—a big enough rise to spur the activity of the sweat glands, which open onto your surface through pores. Perspiration is part of the body's temperature-regulating system, the body's way of cooling itself down. Because the skier has chosen clothing of the right material, perspiration escapes into the air so it can evaporate. Certain fabrics are "breathable": They have openings just the right size to let sweat out, but small enough not to let rain and snow in. You remain comfortable.

2 p.m.

The skier has stopped for lunch, restocking her furnace, and now she's back on the slopes. But the part of you around the foot area feels cold, even though that's the place you're the thickest. Why? It's because she wore cotton socks instead of silk or some breathable synthetic. They're trapping the moisture instead of letting it escape. Uh-oh. Cold feet are the worst!

4 p.m.

But they're not as bad as a cold head! The skier's hat has fallen off. Since humans lose 40 percent of their body heat through their heads, it looks like you're in for a chilly afternoon. The Sun shines brightly against the brilliant white surface of the snow, reflecting light energy. That helps. But still, the part of you in the scalp area tightens and goosebumps begin to form. Because she's female, the skier has a higher percentage of body fat. That gives her a chance of staying warmer a bit longer than a male. But soon muscles all over her body begin to twitch under you. That extra fat can't work miracles. She's shivering. She points her skis downhill and heads for the lodge.

5 p.m.

As the skier walks into the lodge, you start to feel warm again. Why? There's a huge fire blazing, liberating heat. She takes off her ski boots and puts her feet up in front of the fire. As her socks dry out, the part of you around her feet starts to feel nice and warm again. The part of you covering the rest of her body, though, is soon too hot. The skier removes layers until you're completely comfortable all over. At last!

Activity

ROUND THE BEND **Try this experiment to see heat energy in action. Take a jumbo paper clip and hold it in your hand. Notice how warm or cold it feels. Bend it back and forth between your hands. What changes do you notice? Bend it many times. Is there a change in its temperature? Explain your observations.**

The POWER of HEAT

PICTURE THIS

You can see the effects of heat all around you: green grass turned yellow, dried-up bodies of water, hot tar. Whether energy comes from the blazing Sun above or the molten rock beneath, extreme heat affects many areas of the earth. Here are some places (and things) that can—and can't—beat the heat!

Suit Up

Occasionally firefighters must walk through flames to extinguish a fire. Luckily for them, heat-reflective suits (right) have been invented. These suits, made of aluminum and other metals, protect firefighters by reflecting the fire's heat.

When people are exposed to too much sun and heat, their skin may crack and peel. The same thing happens to mud. In extreme heat, water and moisture in the mud dry up and leave behind a cracked, brittle surface like the surface shown above.

GUSH RUSH

Geysers are spouts of boiling water and steam that shoot up from underground. What makes a geyser boil over? Intense heat and pressure. Deep under the surface of the earth, water is heated by magma and turned into steam and extremely hot water, more than 300°F (148°C). All that heat builds up pressure, forcing the superhot water out through a crack in the surface. A geyser's water may be more than 500 years old—about the time it takes groundwater to trickle deep into the earth, become heated, and head back up.

CAKED LAKE

The Mojave Desert in California is one of North America's hottest and driest areas. It's also home to Bristol Dry Lake, a dry lake bed that's caked with 5 feet of salt (above). Thanks to extreme summer heat—100°F (38°C)—water evaporated and left behind millions of tons of salt that had previously been dissolved in the lake water. Today the salt is commercially mined, making its way to dinner tables nationwide.

STEAM BATH

Iceland is one steamy place. It has the most hot springs and volcanic vents in the world. Hot springs (left) contain water that has been heated deep underground by volcanic activity. Icelanders use this hot water year-round to heat their homes and businesses. Pipes several miles long bring hot water from the springs to residential and commercial areas. Some hot springs are for outdoor swimming.

Bbbbrrrr!

Just because a place looks and feels cold doesn't mean there's no heat. Heat is everywhere, even in the Arctic. A research team called SHEBA (an abbreviation for Surface Heat Budget on the Arctic Ocean) is working hard on an eight-year project to measure heat flow in a 39-square mile area of the Arctic. The goal is to see how climate and temperatures in the Arctic affect the rest of the globe.

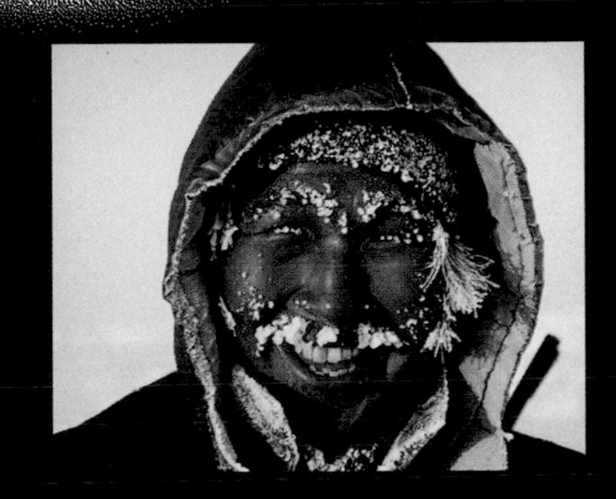

A Picture Tells It All

A house can lose a lot of heat during the winter. All objects have heat and radiate infrared energy. A thermograph detects and measures variations in an object's heat, and converts it into a picture. This image shows just how much goes out the window! (The color black indicates the coldest areas, followed by blue, orange, red, and white. White indicates the hottest areas—where the most heat is lost.)

Activity

HEAT SNAP **Photographers trek the earth to take pictures of extreme weather conditions. You can do the same—just grab a camera. Document the summer heat or cold winter around you by taking pictures of natural phenomena. Look for evidence of extreme weather, such as frozen water or evaporated lakes. Make a hot or cold scrapbook and share your photos with the class.**

HOT SHOT

Something strange had been going on in Uncle Sonny's old barn for some time now. Well, ever since his niece and nephew Summer and Kelvin came for a visit during a school break. Uncle Sonny had stopped using his barn more than five years ago—and no one had stepped foot inside the old structure since. Summer and Kelvin loved visiting Uncle Sonny, who was a professional nature photographer with a special love for animals. He let them experiment with his cameras and films and taught them how to develop the pictures in his darkroom. The barn had been one of Summer's favorite subjects for photographs, but now it was off limits. Nevertheless, Uncle Sonny continued to encourage the kids to learn about infrared photography.

"I don't want you kids to go too close to the barn for a while," Uncle Sonny warned one evening at the dinner table. "I've been hearing strange noises at night—like rustling grass. I think someone—or something—might be sneaking into the barn at night."

Kelvin raised his eyebrows and cast Summer a wary look as he took another bite of his chicken drumstick. It wasn't like Uncle Sonny to be so cautious.

"Be careful, then," said Uncle Sonny, clearing their plates. "I wouldn't want you to come face-to-face with stray animals or a burglar or anything like that." He cut up the leftover chicken in small bits and sealed it in two small plastic bags. "Now I've got to finish up some work in the darkroom." Uncle Sonny disappeared into his photo lab.

"Hey, let's see if we can hear anything in the barn," said Kelvin.

"Well, okay," smiled Summer, "I'll bring a camera—let's see if we can catch something on film!"

"The Moon is new right now, so it's really dark. We might not be able to see anything," observed Kelvin, peering out the window.

"Don't worry," Summer assured. "Uncle Sonny's special infrared film for night shoots should catch almost anything." Infrared film is sensitive to heat, so even in darkness it can record any object that radiates infrared light.

Fifteen minutes later, Summer had the camera loaded, and she and Kelvin left the house. "Let's split up," suggested Summer. "You go around the back of the barn, and I'll go around the front. We'll stake the place out until we hear something. If one of us does, give a short-long-short whistle, and we'll meet back here."

"Agreed," whispered Kelvin.

Kelvin had been listening behind the barn for no more than 10 minutes when he heard rustling noises and high-pitched cries coming from inside the barn. He let out the secret whistle and ran back to the house. Summer followed right behind. "Did you hear that?" Kelvin whispered.

"I heard something, all right," Summer replied. "Almost like a purring sound. And I took a bunch of pictures. I couldn't really see anything, so I don't know how they'll turn out."

The next morning, Summer got up early to develop the film before breakfast.

About an hour later, she emerged from the darkroom. "I think I messed up—this is black and white film, but it doesn't look right." She showed Kelvin the photographs. The photos almost looked like negatives: The trees surrounding the barn showed up with white foliage.

"Look at this!" Kelvin's eyes widened. "I think it's a ghost!" A figure in the shape of a person hovered just inside the barn door.

"Wow, you can even see the veins in the arms!" exclaimed Summer.

"But what are these?" Kelvin asked,

pointing to more white figures in the corner of the barn—small white, puffy ball-like objects with tails.

"Good morning," Uncle Sonny said, coming in the back door with an empty plastic bag. "What have you got here?"

"Well, we got a little curious about the barn—" Summer started.

"—but look," interjected Kelvin. "The barn is haunted! We caught a ghost on film!"

Uncle Sonny studied the photograph. "You took this picture?" he asked, looking at Summer.

"Yes, last night. But I obviously goofed up when I developed the film."

"Actually, these pictures turned out just fine," smiled their uncle. "Infrared film detects slight differences in temperature and can capture things that we can't see as visible light."

"Things like ghosts?" Kelvin asked.

"What do you think?" replied Uncle Sonny.

Was the figure in the picture really a ghost? Use the clues below to figure out the mystery.

Clues

Use these clues ...

1. **Infrared film is sensitive to infrared radiation, light in the invisible portion of the electromagnetic spectrum.**
2. **We experience infrared light in the form of heat.**
3. **Infrared film can record any object that radiates in infrared light—even in pure darkness.**
4. **Organic substances, like vegetation—or people—reflect infrared light more strongly than other substances. Such objects show up as white on developed film.**
5. **Infrared film can even detect heat from human blood, so veins become visible under the skin!**

Answer on page 32

think thermal

This fire won't melt the ice because it's fake.

The Ice Hotel in Sweden is surely the coldest hotel—in temperature that is! Made of snow and ice, its average indoor temperature is about 18°F (-7.8°C)! Every year the owners receive about 7,000 guests, who eat off ice plates, drink from ice cups, and sleep on beds of ice. How do they stay warm? Reindeer skins are placed on the beds and every guest is outfitted with snowsuits, gloves, and hats. But each spring the Sun melts the hotel. The whole place is rebuilt the next year!

Heat Riddle

An **engineer**, a **physicist**, a **mathematician**, and a **mystic** were asked to name the greatest invention of all time.

The **engineer** chose fire, which gave humanity power over matter.

The **physicist** chose the wheel, which gave humanity the power over distance.

The **mathematician** chose the alphabet, which gave humanity power over symbols.

The **mystic** chose the thermos bottle.

"Why a thermos bottle?" the others asked.

"Because the thermos keeps hot liquids hot in winter and cold liquids cold in summer."

"Yes—so what?"

"Think about it," said the **mystic**. "That little bottle—how does it know?"

In the Red

Infrared film, a special type of film that can be used in most cameras, really comes in handy! It is used in law enforcement, medicine, and farming. To solve crimes, forensics technicians use it to detect gunshot-powder burns. Doctors use it to reveal medical conditions invisible to the eye or X-rays. It also shows farmers where insect damage is bad in their crops. **For more on infrared film, see Hot Shot, page 28.**

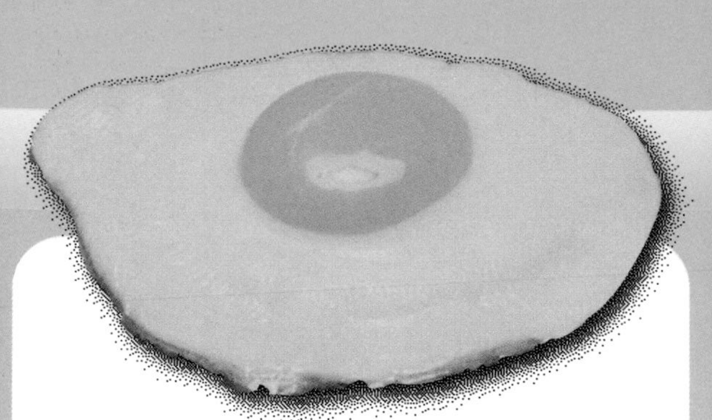

It's No Yolk!

Residents of Oatman, Arizona, fry eggs on the sidewalk every Fourth of July, but they're not making omelettes. They're demonstrating that their city is hot stuff. The contestants use mirrors and tools to reflect the Sun's light. Without reflected heat, the sidewalk would need to measure a steady 110–120°F (43–49°C) to be hot enough to fry an egg.

Firewalkers?

Is a person walking on hot coals using supernatural powers or simply physics? Scientists have studied how heat is conducted from the coals to the feet. During a quick walk, the heat from the coals does not move fast enough to penetrate a protective layer of air and water on the skin. This layer breaks down quickly, so the walker must move fast.

Fun Facts about Heat

1. When you heat water to 212°F (100°C), its temperature stops rising, even though the water is still being heated. Some of the water particles break free and turn into gas. This is boiling.
2. The Concorde, a supersonic jet, flies so fast that the heat caused by flying through the air makes the airplane expand up to 9 inches in length during flight.
3. Understanding that heat makes things expand can help you open a stuck jar lid. Run hot water over the lid; the metal expands faster than the glass jar. Heat helps the top pop.
4. Ice cubes cool drinks because the heat of the drink travels into the cold ice cube. The ice cube heats up, and the drink simultaneously cools down.
5. If you plan to travel in deep space, bring a thermal space suit. In the empty space surrounding the stars, the temperature drops to -454°F (-270°C), just a few degrees above absolute zero, the coldest temperature possible.

YOUR WORLD YOUR TURN

Keeping Warm and Keeping Cool

What is the best way to hold onto heat? What is the best way to remove heat?

Split up into three groups, and have a third of the class research home heating. Go to your local library and research the best ways to heat a house. Consider solar heat, electricity, oil, or a wood-burning stove. As you do your research keep in mind that certain factors, such as insulation, will affect how warm your house stays. Then design an energy-efficient house.

After you've designed your "dream" house, consider the following questions:

- Why is this an efficient or inefficient way to heat a house?
- What are the disadvantages?
- What are the costs involved?
- How could you make this method of home heating better?
- Compare and contrast your home-heating method with your classmates' methods. What are the similarities and what are the differences? Is one kind of home heating better than another?

- Another third of the class should research home cooling. What is the most energy-efficient way to cool a house?
- The last third of the class should research other methods of regulating heat. How is an igloo heated? How is a tropical hut cooled?
- All students should consider how the design and construction of a house affect its temperature.

Ready for the ultimate challenge? Enter this or any other science project in the Discovery Young Scientist Challenge. Visit *discoveryschool.com/dysc* to find out how.

ANSWER Solve-It-Yourself Mystery, pages 28–29

The "ghost" in the picture was really a heat image of Uncle Sonny captured on infrared film. He had discovered a stray cat and her litter of kittens in the barn and delivered chicken scraps to them at night and in the morning. He wanted to keep his discovery a secret until he knew the kittens would be safe to handle. The litter of kittens also showed up as little puffy balls of white on film, the "hottest" things in the barn besides Uncle Sonny.